FIRE!

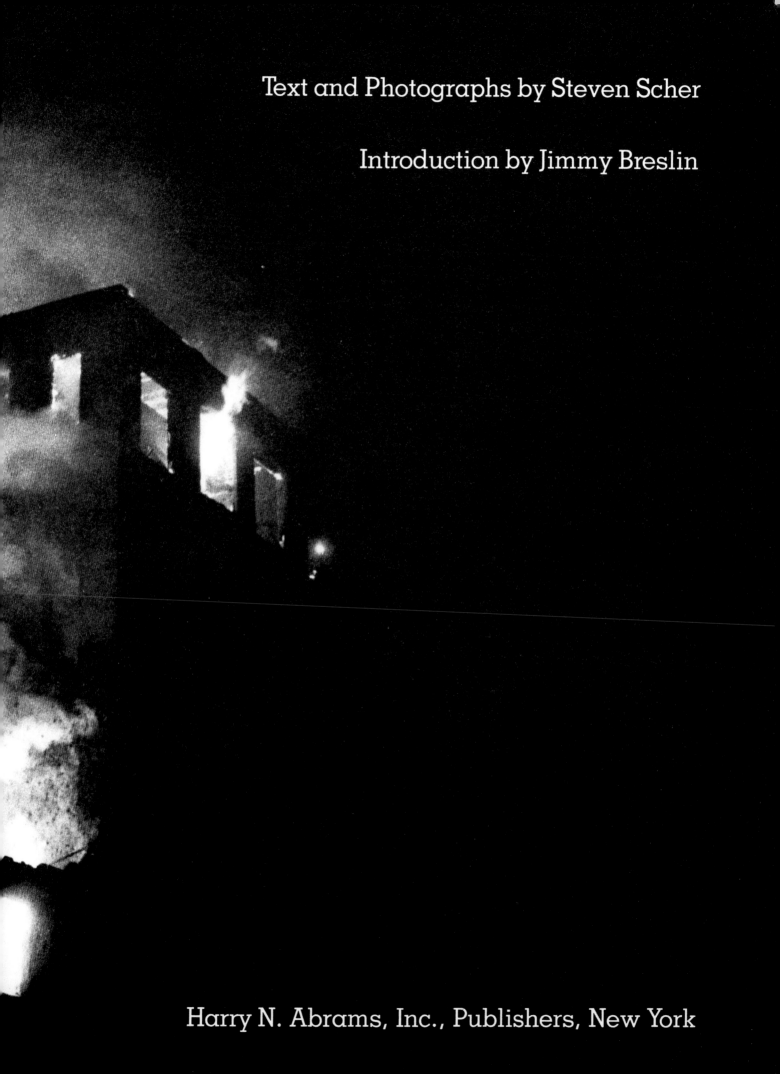

Text and Photographs by Steven Scher

Introduction by Jimmy Breslin

Harry N. Abrams, Inc., Publishers, New York

To the person who gave me my first
camera: my father.

Book Designer: John S. Lynch

Library of Congress Cataloging in Publication Data

Scher, Steven.
    Fire!

    1. New York (City)—Fires and fire prevention.
I. Title.
TH9505.N509S3   628.9'2'09747'1   77-27270
ISBN 0-8109-0905-7
ISBN 0-8109-2151-0 pbk.

Library of Congress Catalogue Card Number:
77-27270

Reprint of the Jimmy Breslin column of May 26,
1977, courtesy of the Chicago Tribune–New York
News Syndicate. Copyright © 1977 by The New
York News, Inc., New York.

Published in 1978 by Harry N. Abrams,
Incorporated, New York. All rights reserved.
No part of the contents of this book may be
reproduced without the written permission of
the publishers

Printed and bound in Japan

CONTENTS

THE ENEMY

FIREHOUSE

WARRIORS AND WEAPONS

ALARM

THE BATTLE

VICTIMS AND ONLOOKERS

THE AFTERMATH

THE VALIANT—
The Victims and The Fallen

TARGETS FOR TOMORROW?

ACKNOWLEDGMENTS

PREFACE

Colonial America had a real romance with firefighting. Volunteer firemen enjoyed a great deal of status throughout the colonies. Having been selected as "able, discreet, sober men," the volunteers were an elite group whose ranks included John Hancock, Paul Revere, Benjamin Franklin, and George Washington. Each volunteer company had a piece of apparatus individually named, painted, funded by a group of financial supporters, and often had ballads written about their exploits.

When the cry "Fire!" echoed along the cobblestone streets of colonial New York, relaying the alarm from block to block, residents of the closely packed wooden houses would throw out the leather buckets they were required by law to keep for this purpose. The buckets would be gathered by the volunteers as they ran to the scene of the fire. Other volunteers would race dozens of blocks pulling fire pumpers that had such affectionate nicknames as Old Brass Backs, The White Ghost, Red Rover. The scene at a fire was one of seeming pandemonium, with bells clanging and people running in all directions.

Much value was placed on getting "first water" on a fire. Competition among the various volunteer companies was fierce, and two companies racing to a fire might easily collide. Brawls and fist fights would then prevent any water from reaching the fire.

Out of this tradition of pride, love of life, bravado, competition and teamwork, machismo and legend, the New York Fire Department was born.

Today, much of the early romance of firefighting is gone. In its place is a war—a war with a multitude of enemies. A New York City fireman's work can kill him at any moment, and his feelings about himself and his life reflect this. Because he knows that he may die at any time he knows that he must fully live every instant, yet he cannot control the passage and wastage of time. He is a tough yet tender man, and he loves his fellow firemen as if they were brothers, yet he must be prepared to accept the fact that each year many of his brothers will be killed or injured. While on a personal level he succeeds in countering destruction, he sees its rate increasing and takes that as a measure of the decline of the society in which he lives. The fireman is very likely to be politically conservative, and sometimes his attitudes may appear to be racist, yet he will not hesitate to risk his life by running into a burning building to save a person he can be sure is of a different color. What the fireman sees as the nobility of his profession transcends lesser, personal feelings about his fellow man.

In what the firefighter sees as a world beset by chaos and destruction, fighting a fire is a noble act pitting good against evil, man against destructive acts of nature. It is a clear issue, a battle with no grays, only black and white. In a world of questionable and nebulous values the fireman's work is clearcut. He sees fire destroying the lives and property of the rich as well as the poor. He knows that the demon called fire sees no class distinctions.

When the fireman gets out of a burning building the smoke he tastes and coughs up, the tears pouring out of his eyes, the heat in his lungs all are evidence to him that the battle is on a deeply personal level, that it truly is *his* battle.

The fireman does his work enthusiastically. He sees the battle against fire not as a defensive but as an offensive one, a battle in which he can control the outcome. Once a fire is extinguished, its "rekindle" is looked upon as almost a personal affront.

He sees the enemy now as not just fire, but as apathy and hypocrisy as well. He sees the public clamoring for better fire protection but doing nothing about false alarms, blocked hydrants, arson, assaults on firemen at fires. He sees the closing of firehouses and the layoffs of firemen as mindless, tragic political maneuvers.

Today's New York City fireman is fighting a war not only against fire, but against public apathy and ignorance, political expediency and the seeming disintegration of his city.

These photographs are scenes from that war.

Steven Scher

INTRODUCTION
by Jimmy Breslin

Her eyes were deep in her face and they said nothing. Black smears ran across her cheeks. She stood there with her hands in the pockets of a brown corduroy car coat. The water from the hoses ran under a pile of wood with nails sticking from them and the water came up against her feet and darkened her flat brown shoes.

She stared at the sign which said "Wonder" and then underneath it was "Dru." The "g" and the "s" had been seared out by the flames. She stared at the sign and at the firemen climbing over the debris in the entranceway and she waited for one of them to come out and tell her that her husband, James Galanaugh, was all right.

Her husband was inside the drugstore, down in the basement, and he was dead in a pile of cinderblocks and tin and wood which had burned over him, and the firemen were digging and trying to get at his body so they could drape it in a gray blanket and carry it outside.

"Maybe he could be alive," Mrs. Galanaugh said.

The firemen standing with her looked down at the water running against their boots.

A big guy in a black raincoat came up to speak to her. He was Gerry Ryan, then the head of the Uniformed Firefighters Association.

"Why don't you go someplace and rest?" he said quietly. "We'll come the minute we know something."

"I want to be with him," she said. "Maybe he'll be all right."

Her hand came out of her coat and she ran it through her uncombed reddish blonde hair.

"My father dies on this job," she said. "My father first. And now my husband. Is my husband dead, too?"

Gerry Ryan stood in front of her and his seamed face broke and he wept while the woman in front of him stood and looked at the place where she thought maybe they would find her husband alive.

Firemen were crowded at the entrance to the drugstore, forming an alley between the store and a red-and-white ambulance which had been backed to the curb.

Then there was a murmur and the firemen were taking off their helmets and three of them came up onto the top of the debris in the doorway. The three were bent over and their hands were holding the front end of a wire basket that held the body of a fireman who was dead and wrapped in a gray blanket.

They came down the pile of debris and the ones holding the back of the basket came after them and they walked through this alley of smeared faces that were wet with tears and they put the body into the ambulance.

And a chaplain, Canon E. J. Downes, his dirty hands clasped in front of him, followed the basket and said, softly, "Under God's gracious mercy and protection we commit you."

A heavy man who looked like her brother, and a fireman with his helmet off in respect, put their hands on Mrs. Galanaugh's back. They guided her away from the water and the hoses and the store her husband died in. Her feet dragged as she moved. She did not cry. She looked straight ahead and around her were the sounds of the business her father and her husband were in.

"Tommy Reilly went right in, he wasn't in there a minute and he was gone."

"How many is that now, six or seven out?"

"Al Hay was next door. He felt this shock wave and he got thrown against the wall. He knew it was bad then."

"Where's the chief? We're not supposed to do anything until the chief tells us."

They guided her across the street, to the sidewalk in front of Madison Square Park, and Mrs. James Galanaugh walked under the trees in the bright fall sunlight and everybody looked at her and nobody spoke while she passed.

She had lost her husband fighting fire in this city. And there were eleven others who had died in this fire on Twenty-third Street. Twelve dead firemen. Twelve widows.

Gerry Ryan stood in the middle of the street. He closed his eyes. "There are fifty-one children involved in this," he said. "Fifty-one children." Somebody standing next to him started to throw up.

For Engine Company 18, it was no good right from the start. At 9:36 P.M., when the bell began ringing in the old, narrow firehouse, the fireman on watch listened as the bell hit five times, then nine times, then eight times. Box 598. The fireman didn't like it at all.

"That's a bad box," he said.

The fireman went to a card in front of him. The card said that for fire alarm box 598, Engine Company 18 is first due on a second alarm.

Monsignor Stanislaus Jablonski, a Fire Department chaplain, was standing with the fireman.

"That box has been cooking for a long time," the fireman said. "You're going to see a second alarm for it."

At 10:05 P.M. the bells rang again. First two quick rings. Then two more quick rings. Second alarm. Then the bell began tapping out the numbers 598. And the firemen in the house stepped out of their unlaced shoes, put on their boots, climbed up on the noisy Mack pumper and were still putting on their turnout coats when the engine came out of the house and onto West Tenth Street. It turned east and headed for box 598, which was for Broadway and Twenty-second Street. When they got there, four of them would die so quickly that it would take time for anybody to realize it.

The fire burned all through that Monday night and into the daylight and it took the lives of twelve firemen. It did not take them in one fast, complete motion. Firemen do not die easily. They die in blackness on basement floors and they die with cinderblocks crushing their bodies and burning debris eating through their coats at their bodies and they lie there with the air in their masks running out and their bodies are on fire and they scratch at the cement floor and they try to get away and they strangle to death and the flames at their bodies leave nothing.

They sit around a lot, these firemen of New York. They sit and they complain and they become narrow-minded and they are so much above it all because of the way they live when they work. When they work, they come in a matter of multiples of seconds and they never back off and they come against all the statistics.

For fifty years now, an average of one fireman each five weeks has died in the line of duty. And one of every five firemen each year suffers a permanent injury. One stomach full of smoke is enough to wreck a body forever, and firemen live an average of only eight years after they retire. And they all have large families and the families must suffer too, and on that cool day in 1966, they had the families in the vault room of a bank up at the corner and a little girl sat on a fireman's lap and she kept saying, "See my blue shoes. My daddy bought me my blue shoes. Aren't they pretty?" And her father was across the street dead in the fire.

But still they come, and they come with the pureness of people who are ready to risk their lives for somebody else and you don't want any more out of human beings than this, and in this city we argue over budgets while the firemen die.

Die and leave people. Ben Messina of the Uniformed Firefighters Association came up to Bernard Tepper's split-level house in Baisley Park, Queens, that night in 1966 to talk to the widow about funeral arrangements. A nine-year-old boy, one of Tepper's three children, was on the stoop.

"Are you a fireman?" the boy said.

"Yes, I am," Messina said.

"Were you saved?" the boy said.

Messina didn't answer.

"My daddy wasn't," the boy said.

On West Tenth Street, the purple-and-black bunting blew in the light rain falling on the front of Engine Company 18's house. Edmund Askland, the lieutenant in charge, was inside. He spoke quietly about what had happened to his company.

"They weren't in there more than a minute," he said. "They were trying to find the source of the fire and the floor caved in."

"Why did they have to go in there in the first place?" he was asked.

"You have to go inside," Askland said.

"Why not stay outside and pour water?"

"A basement fire," Askland said. "How are you going to fight a basement fire from the street? You've got to go to the basement. If you don't, we'd have fires burning out of control for days."

A red-haired fireman standing by the door said, "They didn't want to risk their lives any more than I do. But it's the job. They had to go in. It's just a thing that happened on the job."

"Berry," another fireman said. "You hear about Berry from Ladder

Seven? All that was left was the metal tag inside his helmet. And only the 'ERRY' on the tag. That, and the metal wire that goes around inside. The rest of the helmet was gone. Disintegrated. That's case-hardened leather and the fire ate it up. You figure out what was left of Berry himself."

The small door at one side of the building opened and a man carrying an umbrella came in.

"Yes?" one of the firemen said.

"I was just passing by, I just came from out of town and I thought I'd say hello to a relative of mine. Bernard Tepper. Is he here this evening?"

"Tepper?" Askland said.

"Yes, Tepper," the man with the umbrella said. "My name is Jesse McCarroll."

The red-haired fireman looked away.

"Tell him," somebody said.

"Tepper is dead," a fireman next to Askland said. "He died in a fire last night."

"Yes, he died last night."

The man's mouth opened and he turned around and went back through the door and onto the wet street.

The firemen in the house said very little when he left.

The shift that came on at 6:00 P.M. at Engine 18 was missing men because they had died the night before, but the ones who were there had the first call of the night at 6:52 P.M. It came from box 532, which is at the corner of Hudson and Bank Streets. They went out onto Tenth Street and answered the call just as they had answered the call the night before, and what was in their minds as they went through the streets was a thing that they always are going to have to face.

And when Engine Company 18 got to fire alarm box 532, they found somebody had transmitted a false alarm.

The lieutenant looked up when Lenny Smit walked into the firehouse to start his tour.

"I got a letter for you," the lieutenant said.

Smit didn't want to touch the letter. But the lieutenant put it into his hands and Smit, walking up the stairs to his locker, opened it.

"Mayor Beame and I express our sincere regrets," the letter began. After working a year and a half as a New York City fireman, Smit was being laid off. He stared out a window at the hot West Side street. This was in June of 1975 and Lenny Smit had no idea of what he would do for a living. Once he had taught in a parochial elementary school and had tended bar on the side. He couldn't even do that now; school was closed. He went to the phone and called his wife Elizabeth. She handled this particular call all right.

Three days later, the mailman came to Smit's house in Staten Island with another letter.

"Maybe they don't have to lay you off," his wife said.

"Just give me one break here," Smit said. When he opened it, he found a copy of the letter he had received at the firehouse.

On his last day of work, on July 1, 1975, Smit and his wife went to the firehouse to clean out his locker. As his wife began taking things out of the locker, she began to cry. Smit told her to stop. Then he started to cry, too. He was young and somebody sitting in an office someplace with ledgers was taking his way of life away from him.

When he got up the next morning, he had to find a job. He delivered new cars from New Jersey for eight dollars a car. In the fall, his wife's cousin, a principal of a parochial school in Manhattan, got him a couple of days of teaching each week. Then the cousin left. Smit went on unemployment at $95 a week. He wanted to save lives for a living and he was home watching television.

Finally, he caught on as a driver for United Parcel Service. The job paid $225 a week. Walking into work in the morning was not the same. Lenny loved going into a firehouse, but the delivery truck job was there, a rail on a wet deck, and there now was a baby at home. He was going to stay with this job. Then one morning he showed up for work and found pickets in front of the garage.

"You can go home," the boss told Lenny. "They're shuttin' the place down until the strike's over."

He went into the phone booth and called his wife. If we can't keep this job, she said, what chance do we have with anything? She became upset over bills. They just bought a used car and the payments, $68 a month, now were going to be too much.

Smit stood in the phone booth and he tried to calm his wife and he remembered looking out at the garage, filled with motionless trucks, and he could get a purchase on nothing.

He was 27 and he was 5-feet-8 and weighed 165 pounds and his blue eyes were alert. The thing that people who make up budgets do not understand, as they sit at their desks, is that one of the things that you need first in a city is people who fight, which is a business only for the young. In fighting, experience is hardly a substitute for young legs and reflexes that seem to race along the surface of the skin like a loose current. In the name of debt, and when they should have been firing office workers, those old men in charge of New York City were throwing away young people like Lenny Smit, who is a fighter.

To walk the streets and ask for work, and then have to return home and stare at television, can do physical damage to a man every bit as real as a broken bone. It was close, in this case. As Lenny Smit's life started to drip away on sidewalks, common sense made itself felt to the people running the city. How can you lay off firemen in a city that burns down? Fiscal responsibility becomes moral irresponsibility.

In the fall, on unemployment, Smit heard from the fire union that there was a chance that he would be rehired. In November, he received a letter from the city saying he could have his job back. On the day after Christmas, Lenny Smit, his body swinging, walked into Ladder Company 24 on West Thirty-first Street. He had been gone for a year and a half and we nearly had lost him. But now he was back again.

Five months later he was at the firehouse when the voice alarm announced a fire on Twenty-eighth Street, between Broadway and Sixth Avenue, a big fire. The block is in the flower market. By day it is crowded with crates of plants and rows of potted bushes filling the sidewalks; a city street that smells like a country field. At night, the district looks like another empty factory block. The gloomy three-story building at No. 28 West is 100 years old and had been a health club since the 1920s. For the last couple of years it had been called the Everard Baths, an all-night place frequented by homosexuals.

As the truck carrying Lenny Smit came up to the building, there were windows on the third floor, small rectangles, and they were filled with faces.

Get them, Smit told himself. Get them.

His body strained to run and, as the truck stopped, Lenny Smit was off it, hands grabbing for a ladder. He and an older fireman named Loughlin threw one ladder up to the rectangular windows high up on the building. Loughlin went up first and Smit stayed at the bottom, steadying the ladder. Loughlin began pulling a guy through the small window. The truck's aerial ladder came up to the windows. The aerial had a platform. Loughlin got the one guy onto the platform and now Lenny Smit went up the ladder. He went up very fast, and the ladder shook as he came into the smoke and heat and the faces at the window.

"We got you, don't worry about it," Smit called to them. The faces did not answer. Fear had turned them to marble.

Smit reached in and grabbed one guy under the arms and pulled him through the window and hauled him to the platform of the aerial ladder. He turned and went back to the window. Nobody was there. The ones who had been there had passed out in the smoke. Smit and Loughlin had no masks on. They went through the window head first and tumbled onto the floor in the smoke and began to pick up bodies.

Smit remembers that he had his hands on five people, who were pushed and tugged through the windows. He grabbed breaths through the windows and then put his face back into the smoke again. Now they were left with one guy and Loughlin was holding the guy's face up to the window, trying to give him air, and then he pushed him out through the window and onto the aerial ladder. Lenny Smit went down on all fours, with his nose on the floor, and he began to crawl away from the windows. He wanted to get more people.

He kept telling himself that the windows were directly behind him. He

could see no fire, but he could hear it everywhere. Ahead of him, straight back into the blackness, men were screaming. Smit crawled, and his ears began to pain him. The ears tell it to you first: get out. Smit turned around, nose directly against the floor, and began to move like a dog. When he reached the wall, he came up and dove for the window. The ladder was right there.

Altogether, twenty lives were saved by Smit and three or four other firemen who made it to the windows.

When Smit got down to the truck he strapped on a mask and air tank and went into the building through the ground floor doorway. He made his way up to the second floor, but now it was useless. The third floor had collapsed. The water cascading down on Smit was boiling hot. The fire was conquering the water being thrown on it.

Lenny Smit had to turn around and crawl out of the building. He went onto the street and took off his mask and sat down on the truck with Loughlin. They looked up at the windows, which were filled with flame now, and both started to become emotional. They had been forced to leave people inside a fire.

Somebody in charge passed by the truck and he waved for a doctor. The doctor took a look at the older one, Loughlin, put an arm on his shoulder and took him away to the Department's medical office for a complete examination. You worry about them when they're over thirty-five.

The doctor left Smit alone. He knew that Smit would get over it. He was only twenty-eight, which is the perfect age for a fighter. Besides, there is nothing the matter with people like this being upset. It is just another sign of how good they are.

THE ENEMY

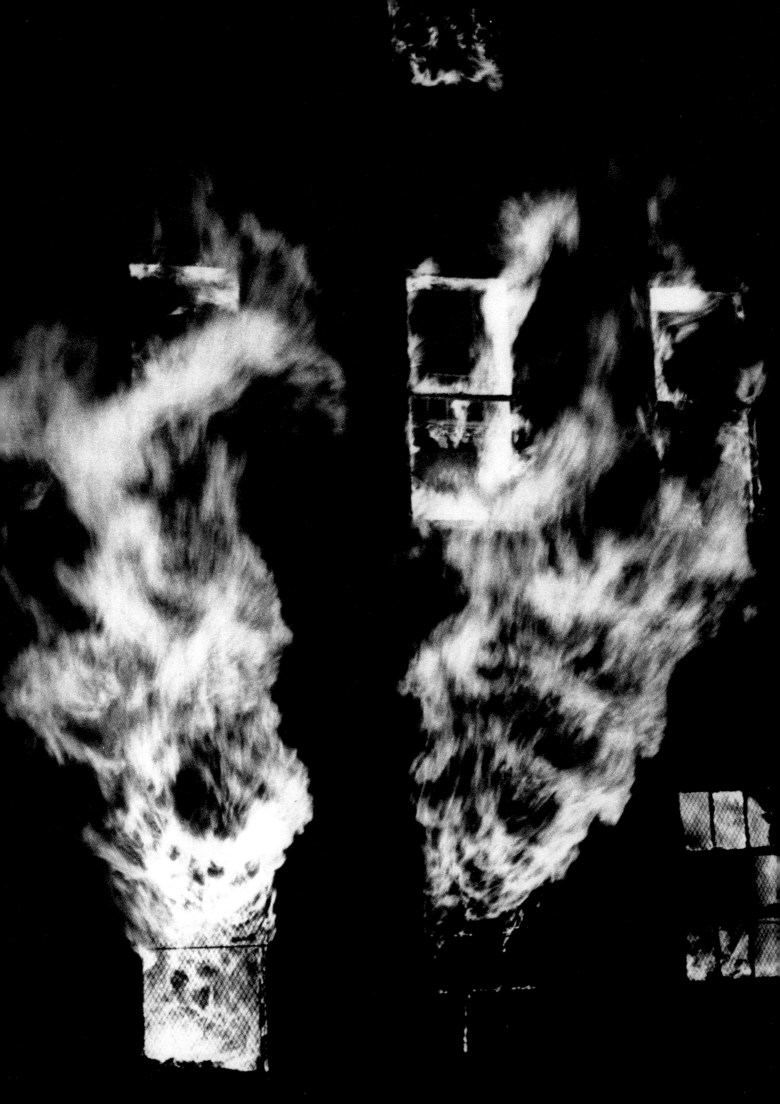

FIREHOUSE

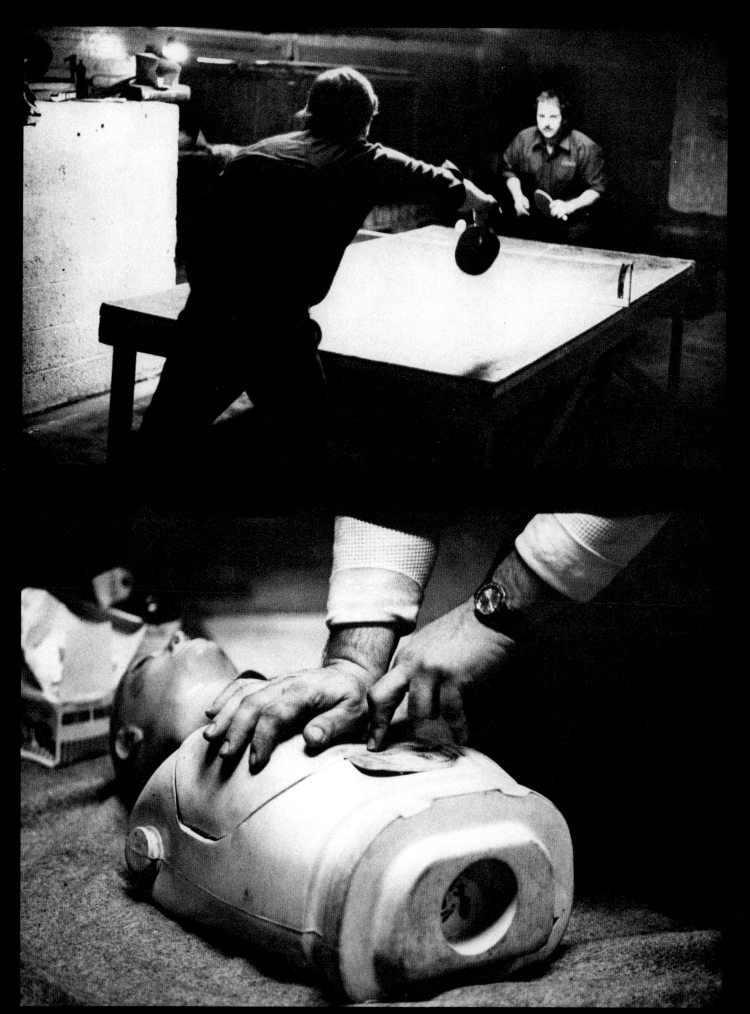
Fireman training in the use of cardio-pulmonary resuscitation (CPR)

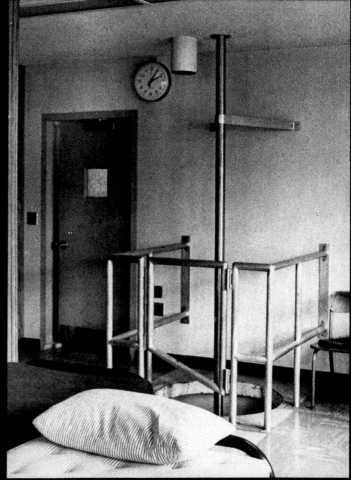

(facing page) Fireman practicing rescue-by-rope techniques

WARRIORS AND WEAPONS

PORT SYSTEM		STBD. SYSTEM	
STANDBY 	EMERGENCY ENGINE STOP 	STANDBY	EMERGENCY ENGINE STOP 
P.H. ALARM TEST 	MAIN ENGINE ALARM 	BILGE PUMP RUNNING 	MAIN ENGINE ALARM 
ALARM ACKNOWLEDGE 	ENGINE CONTROL LOST 	BILGE PUMP OVRD OFF ON 	ENGINE CONTROL LOST 
ENGINEER'S CALL 	HIGH EXHAUST TEMP. 	HIGH BILGE ALARM 	HIGH EXHAUST TEMP. 
GEN. NO. 2 OFF ON 	STEERING GEAR FAILURE 	GEN. NO. 1 OFF ON 	STEERING GEAR FAILURE 
GEN. NO. 2 AVAILABLE 	GEN. NO. 2 GROUP ALARMS 	GEN. NO. 1 AVAILABLE 	GEN. NO. 1 GROUP ALARMS 
PORT PUMP ENGINE ALARMS 		FIRE ALARMS 	STBD. PUMP ENGINE ALARMS 

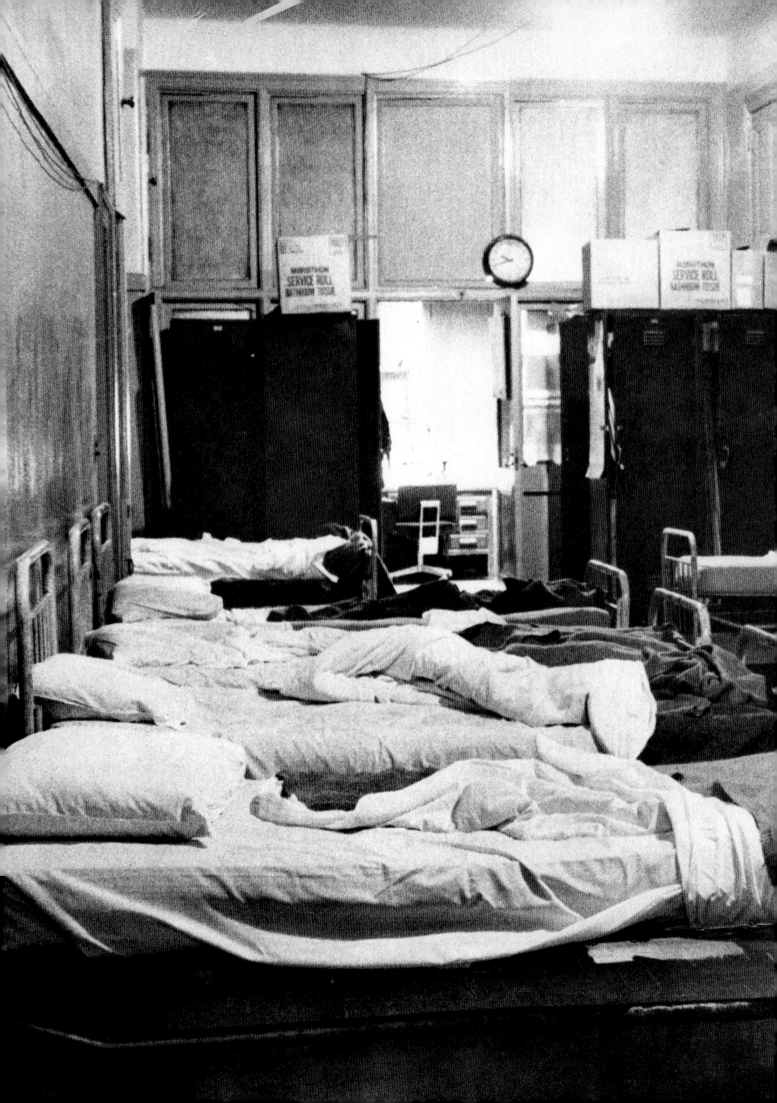

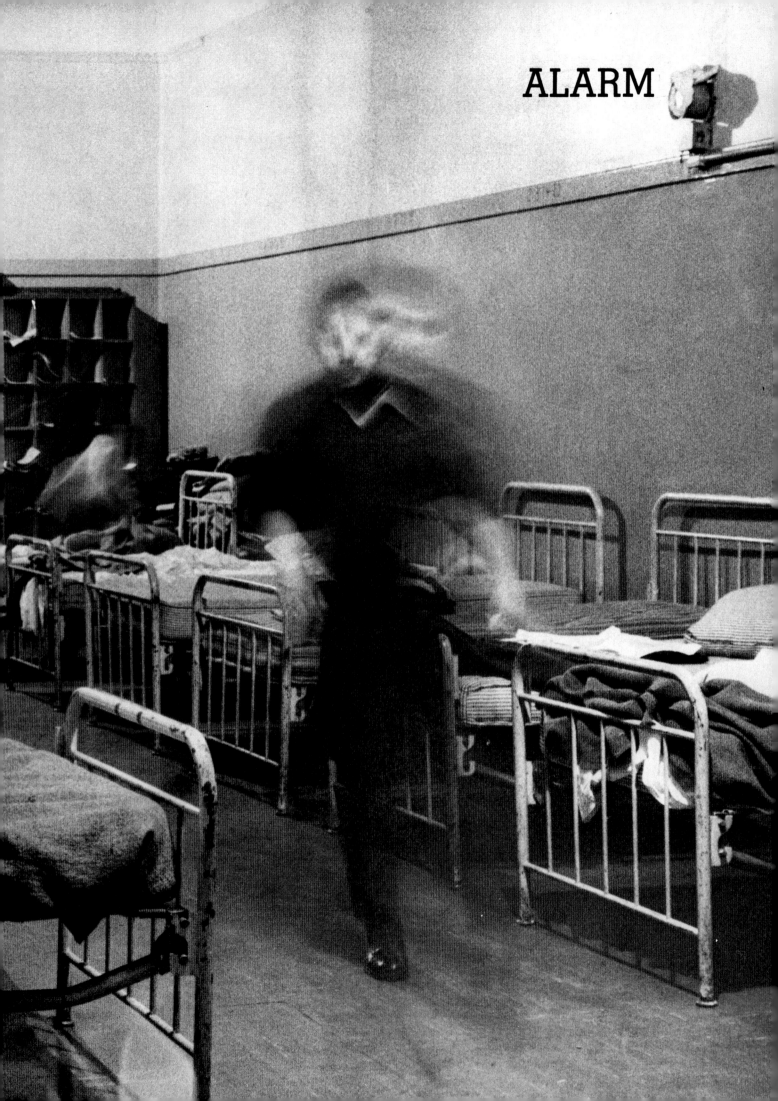
ALARM

Power supplies in central communications office for alarm boxes

Inset shows computer readout screen displaying active fire incidents. Time of photo is 1 A.M. Screen shows alarms being processed at rate of 56 per hour, as well as availability of fire apparatus in Brooklyn

THE BATTLE

OPEN ▶

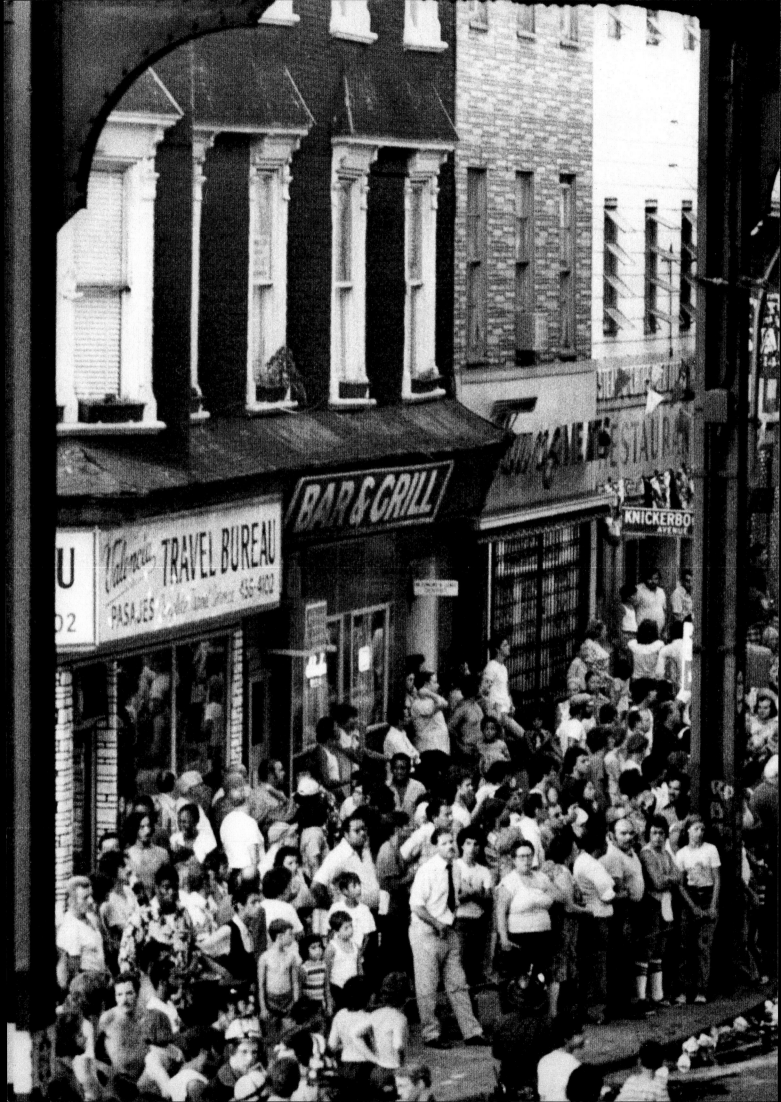

VICTIMS AND ONLOOKERS

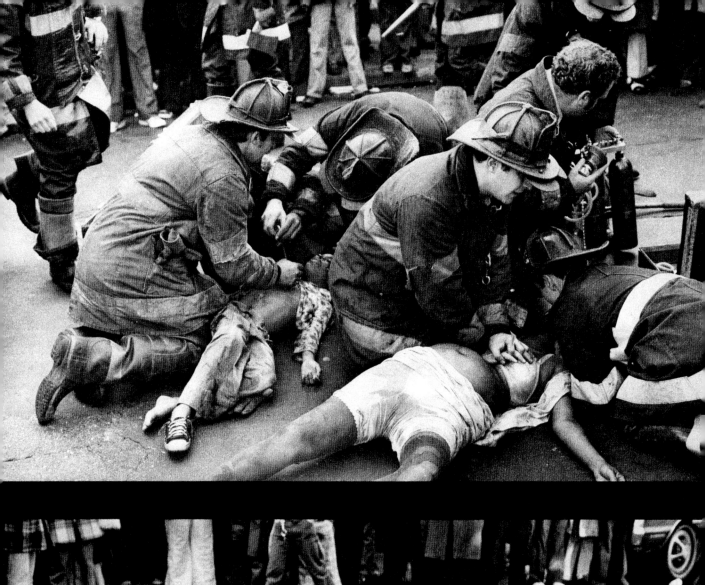

THE AFTERMATH

Fireman searching debris of Everard Baths fire for victims.

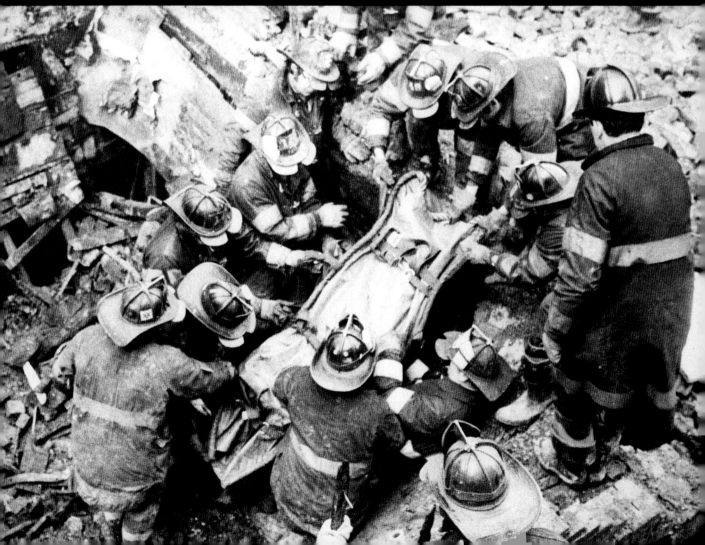

THE VALIANT—
The Victors and The Fallen

MEMBERS OF THE NEW [YORK FIRE DEPARTMENT]
WHO DIED IN THE PER[FORMANCE OF DUTY]

REQUIESCAT [IN PACE]

(following spread) Sign on vacant building literally meaning "All goodies out." This tells potential looters, vandals, and arsonists that nothing of value remains in building

ACKNOWLEDGMENTS:

I wish to thank Diana Goldin, Margaret Kaplan, Robert Morton, and Sandi Nero. Without their confidence in my work, *FIRE!* would not have been possible.